EVERYDAY MYSTERIES

WHAT MAKES A RAINBOW APPEAR?

By John O'Mara

Gareth Stevens
PUBLISHING

Please visit our website, www.garethstevens.com. For a free color catalog of all our high-quality books, call toll free 1-800-542-2595 or fax 1-877-542-2596.

Cataloging-in-Publication Data

Names: O'Mara, John.
Title: What makes a rainbow appear? / John O'Mara.
Description: New York : Gareth Stevens Publishing, 2021. | Series: Everyday mysteries | Includes glossary and index.
Identifiers: ISBN 9781538256558 (pbk.) | ISBN 9781538256572 (library bound) | ISBN 9781538256565 (6 pack)
Subjects: LCSH: Rainbows–Juvenile literature. | Prisms–Juvenile literature. | Meteorological optics–Juvenile literature.
Classification: LCC QC976.R2 O43 2021 | DDC 551.567–dc23

Published in 2021 by
Gareth Stevens Publishing
111 East 14th Street, Suite 349
New York, NY 10003

Editor: Therese Shea

Photo credits: Cover, p. 1 Richard Cassidy/Shutterstock.com; pp. 3–24 (background) Natutik/Shutterstock.com; p. 5 Aluca69/Shutterstock.com; p. 7 Mike Mareen/Shutterstock.com; pp. 9, 13 Nasky/Shutterstock.com; p. 11 Milagli/Shutterstock.com; p. 15 Felix Nendzig/Shutterstock.com; p. 17 Fouad A. Saad/Shutterstock.com; p. 19 (lunar rainbow) CalvinBradshaw/Wikimedia; p. 19 (sunrise rainbow) Mike Pellinni/Shutterstock.com; p. 21 patpitchaya/Shutterstock.com.

Printed in the United States of America

CPSIA compliance information: Batch #CS20GS: For further information contact Gareth Stevens, New York, New York at 1-800-542-2595.

CONTENTS

Look Up! 4
Colorful Curve 6
Where the Colors Come From 10
Double Rainbow 14
Just Right 16
Different Rainbows 18
Make a Rainbow! 20
Glossary 22
For More Information 23
Index 24

Boldface words appear in the glossary.

Look Up!

There are only a few things that make people stop what they're doing and look up. A rainbow is one of them. This colorful sight only appears when **conditions** are just right. Let's find out how a rainbow happens!

5

Colorful Curve

A rainbow often looks like a curve, or a half circle, in the sky. It's made up of many different colors. The colors appear in a certain order. From outside to inside, they are red, orange, yellow, green, blue, **indigo**, and violet.

Here is one way a rainbow forms. Sunlight shines in one part of the sky, and rain is in the **opposite** part. When light passes through the raindrops, it creates a rainbow. A person must be between the sun and the rain to see it.

raindrops
sunlight

Where the Colors Come From

Light is made up of different colors that travel like waves. Each color we can see has a different **wavelength**. Red has the longest wavelength. Violet has the shortest wavelength. The wavelengths of other colors measure between these two.

COLOR WAVELENGTHS

nm = nanometers
(measurements are not exact)

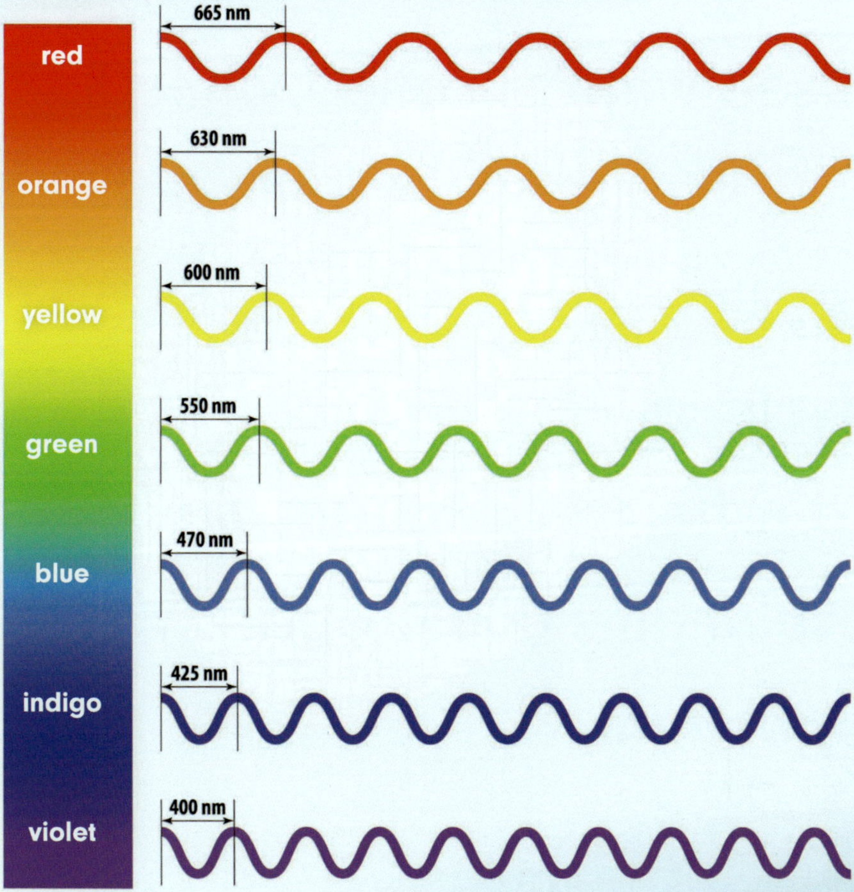

When all colors travel together, we see what is called white light, like sunlight. When white light goes through raindrops, different wavelengths refract, or bend, and **reflect** in different ways. The wavelengths come apart. We see a **spectrum**.

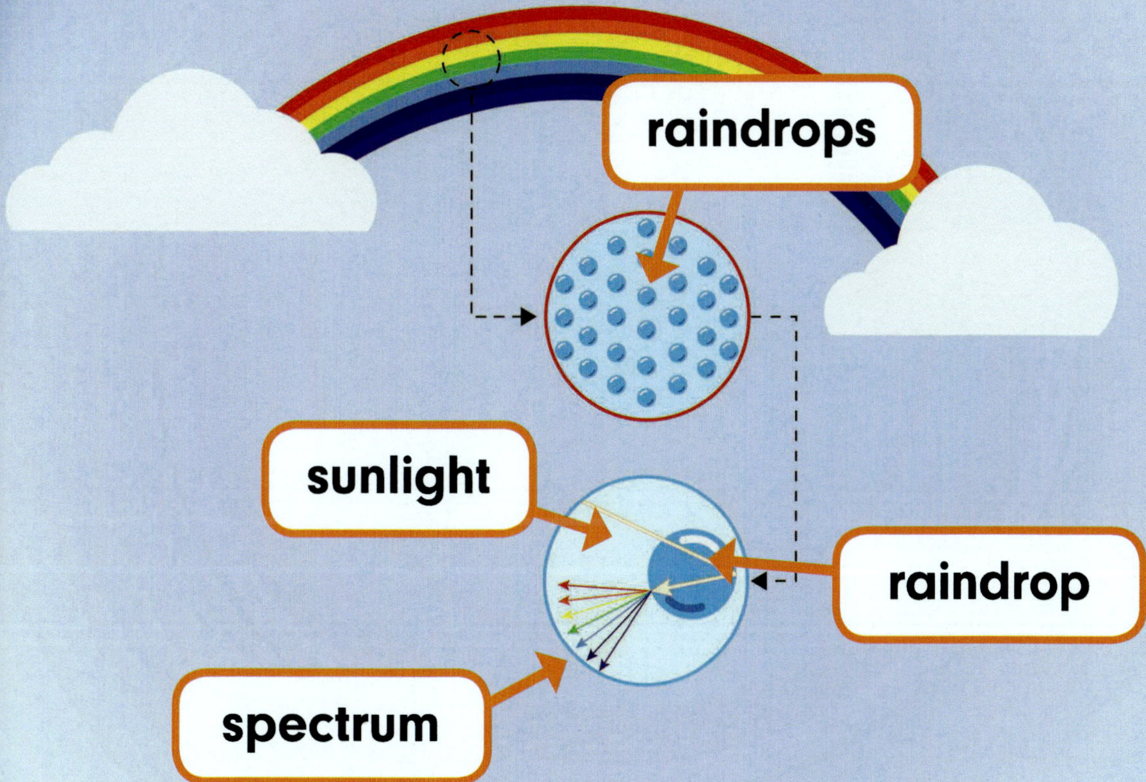
raindrops
sunlight
raindrop
spectrum

Double Rainbow

Sometimes, sunlight reflects two times in the back of raindrops. This can make a second rainbow appear outside the brighter rainbow, or primary rainbow. The colors in the secondary rainbow appear in the opposite order as the primary rainbow.

Just Right

Conditions need to be just right for rainbows to appear. The sun needs to be low in the sky, such as in the morning or late afternoon. The sunlight needs to hit the water drops in the air at a certain **angle** too.

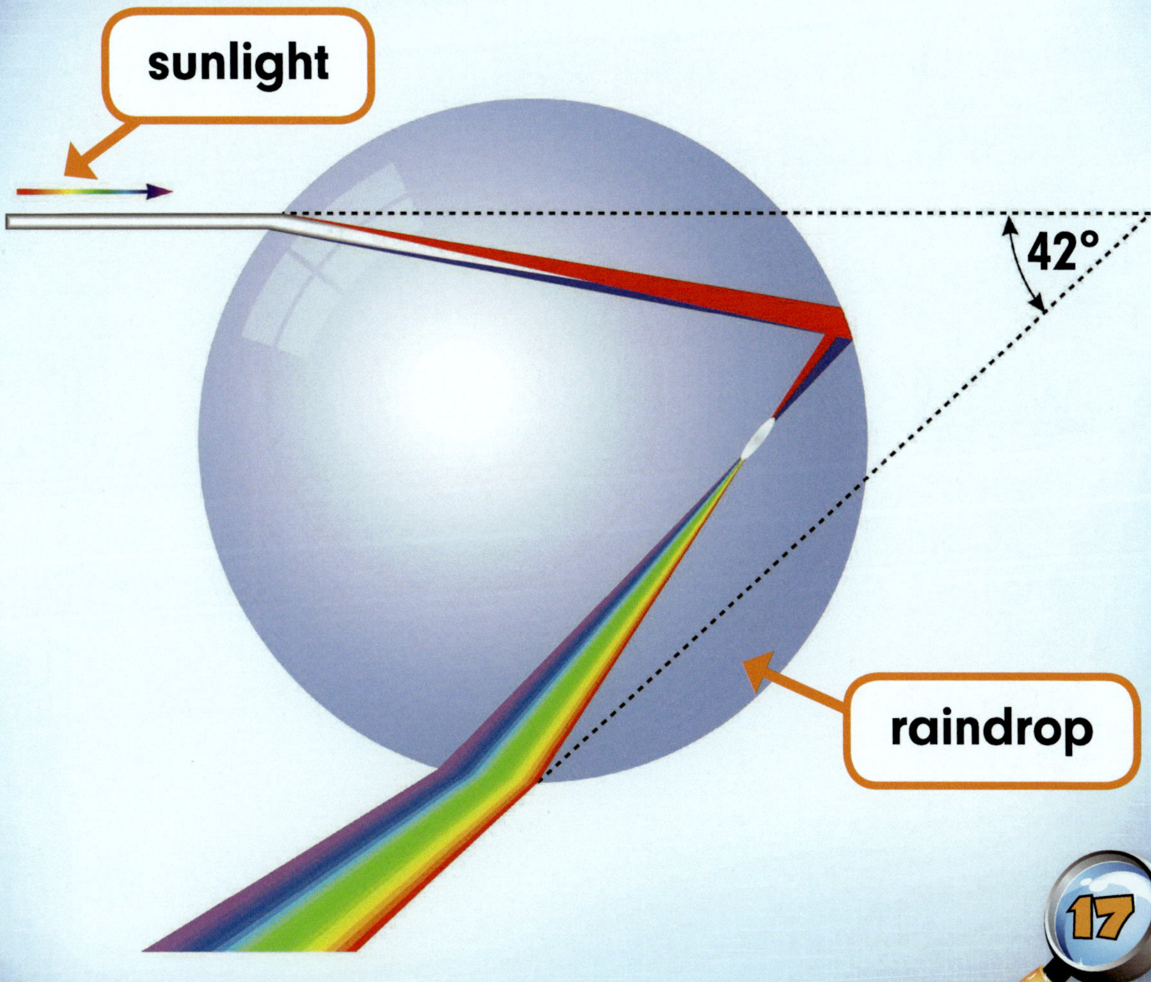
sunlight
42°
raindrop

Different Rainbows

Rainbows don't need rain. Sunlight and fog can make rainbows called fogbows! Fog is made up of bits of water in the air. Sometimes moonlight can make a rainbow, called a **lunar** rainbow or a moonbow. The moon needs to be bright and full.

fogbow
lunar rainbow

Make a Rainbow!

You don't have to wait for rain or fog to see a rainbow. Light shining through a **prism** breaks into a spectrum. You can make a rainbow if you spray water outside on a sunny day too. Try it!

SPOT A RAINBOW!

1. Rainbows are most common in the early morning and late afternoon.
2. Look toward a spot where it's raining.
3. Behind you, the sun shines through the raindrops.
4. The sunlight bends and reflects within the raindrops.
5. The light reflects back to you as a spectrum of colors.

GLOSSARY

angle: the space formed when two lines meet each other

conditions: the things that affect the way something happens

indigo: a deep purple color

lunar: having to do with the moon

opposite: going the other way

prism: a see-through glass or plastic object that separates light passing through it into different colors

reflect: to give back light

spectrum: the group of colors that a ray of light can be separated into

wavelength: the distance from one wave of energy to another

FOR MORE INFORMATION

BOOKS

Slingerland, Janet. *What Causes a Rainbow?* Mankato, MN: Child's World, 2016.

Vilardi, Debbie. *What Makes a Rainbow?* Minneapolis, MN: Cody Koala, 2019.

WEBSITES

How Rare Is a Moonbow?
www.wonderopolis.org/wonder/how-rare-is-a-moonbow
Read more about moonbows.

Make Your Own Rainbow
www.sciencekids.co.nz/experiments/makearainbow.html
Check out this experiment!

Publisher's note to educators and parents: Our editors have carefully reviewed these websites to ensure that they are suitable for students. Many websites change frequently, however, and we cannot guarantee that a site's future contents will continue to meet our high standards of quality and educational value. Be advised that students should be closely supervised whenever they access the internet.

INDEX

angle 16

bend 12

colors 4, 6, 10, 12, 14

conditions 4, 16

curve 6

fog 18, 20

fogbows 18

moon 18

order 6, 14

prism 20

rain 8, 18, 20

raindrops 8, 12, 14

red 6, 10

secondary rainbow 14

sky 6, 8, 16

spectrum 12, 20

sunlight 8, 12, 16, 18

violet 6, 10

wavelength 10, 12

white light 12